Tractors

This North American edition first published in 1986 by
Gareth Stevens, Inc.
7221 West Green Tree Road   Milwaukee, Wisconsin 53223, USA
U.S. edition, this format, copyright © 1986
Text and additional illustrations copyright © 1986 by Gareth Stevens, Inc.
Illustrations copyright © 1985 by Frances Lincoln Ltd.

Conceived, designed, and first produced in the United Kingdom with an original text copyright by Frances Lincoln Ltd.

Library of Congress Cataloging-in-Publication Data

Thompson, Graham, 1940-
    Tractors.

    (Wheels)
    Includes index.
    Summary: Describes tractors and depicts people operating them in a rural setting.
    1. Tractors—Juvenile literature. [1. Tractors] I. Title. II. Series: Thompson, Graham, 1940-
Wheels.
TL233.T48   1986   631.3′72   86-5689

ISBN 1-55532-127-5
ISBN 1-55532-102-X (lib. bdg.)

Art direction and design:
    Debbie MacKinnon & Gary Moseley
Additional illustration/design: Laurie Shock

Typeset by: Ries Graphics, ltd.
Series editor: Mark J. Sachner

Tractors

Graham Thompson

Gareth Stevens Publishing
Milwaukee

Plow

Plows break up hard soil. They also leave long rows. These rows are called furrows.

4

Disk Harrow

Harrows have sharp wheels.
They break up the soil even
more. Now the soil is ready for
planting.

5

Seed Planter

This seed drill drops many seeds into the earth. The farmer wants them to grow into crops. But birds like seeds, too!

Hay Loader

This tractor has a long arm. The arm loads hay into a trailer.

9

Vineyard Tractor

Grapes grow in vineyards on vines. This tractor cuts weeds away from the vines. Why is it so high off the ground?

Crop Sprayer

Sometimes insects eat crops.
Spraying kills insects and saves
crops. This makes farmers happy!

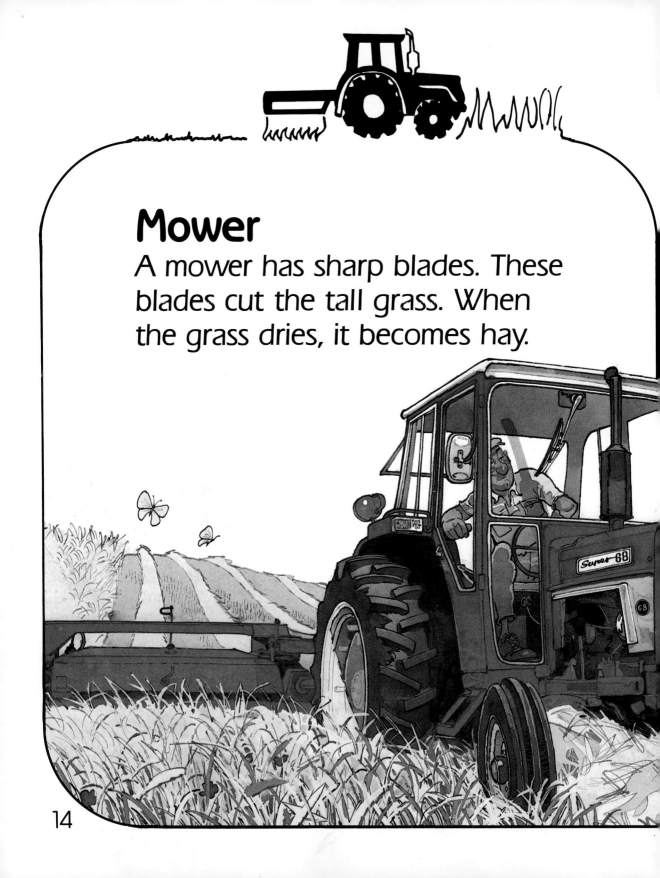

Mower

A mower has sharp blades. These blades cut the tall grass. When the grass dries, it becomes hay.

Baler

The baler picks up the hay. It packs the hay into bales and drops them behind.

Bale Stacker

A trailer takes the bales to a barn. The fork lift stacks them inside. This will keep them dry.

16

17

Grapple Skidder and Log Handler

These tractors work in the woods. The grapple skidder has claws. They lift logs onto the log pile. The log handler has jaws. They can carry many logs.

Combine Harvester

The crop is ripe and ready for cutting. Now the combine goes to work.

It cuts the stalks and collects grain. Soon it is full. Now it pours the grain into a truck.

Snowplow

A snowplow to the rescue! Now the roads will be clear and safe.

Index of Tractors